Forestry Commission

Practice Guide

# Managing and controlling invasive rhododendron

Colin Edwards

Forestry Commission: Edinburgh

First published in 2006 by the Forestry Commission
231 Corstorphine Road, Edinburgh EH12 7AT.

ISBN 0 85538 704 1

EDWARDS, C. (2006).
*Managing and controlling invasive rhododendron.*
Forestry Commission Practice Guide.
Forestry Commission, Edinburgh. i–iv + 1–36 pp.

Keywords: herbicides, invasive rhododendron, rhododendron control, *Rhododendron ponticum.*

Printed in the United Kingdom
on Robert Horne Hello Matt

FCPG017/FC-GB(ECD)/IA-2K/SEP06

Enquiries relating to this publication should be addressed to:

Forestry Commission
231 Corstorphine Road
Edinburgh
EH12 7AT

T: 0131 334 0303
F: 0131 316 4344
E: publications@forestry.gsi.gov.uk

The author may be contacted at:

Forest Research
Northern Research Station
Roslin
Midlothian
EH25 9SY

T: 0131 445 2176
F: 0131 445 5124
E: colin.edwards@forestry.gsi.gov.uk

## Acknowledgements

The field experiments using herbicides were managed by the Technical Support Unit of Forest Research, especially D. Tracy. Bush growth and dispersal experiments were organised by D. Kohn, J. Travis, C. Stephenson and R. Atkinson.

## Disclaimer

The list of products referred to in the tables of this Practice Guide is not comprehensive; other manufacturers may be able to provide products with equivalent characteristics. Reference to a particular manufacturer or product does not imply endorsement or recommendation of that manufacturer or product by the Forestry Commission.

# Contents

# Introduction

This Practice Guide sets out the sequence of events required to plan and manage the control of the invasive plant species *Rhododendron ponticum* (Figure 1). In addition to providing guidance on planning and a decision support flowchart, this Guide presents a number of approved control techniques that can be used individually or in combination to achieve the objectives of rhododendron bush management or eradication for any rhododendron infested site.

In this Guide the problematic invasive wild type of *Rhododendron ponticum* is referred to simply as 'rhododendron', while particular species are given their full botanical name, and other members of the genus in general are described as '*Rhododendron* species'.

**Figure 1** Rhododendron invading heath and young woodland in Kincardine, Scotland.

The initial steps in any control programme should be to identify and quantify the extent of the problem, then choose the most efficient and appropriate method of control within the constraints of site and resource availability (see Choosing appropriate control treatments, page 15). Long-term planning of operations and the commitment of expenditure are required to ensure the work is completed successfully and the objectives are met, for example, eradication within an entire woodland area may take 7–10 years.

## History of introduction into the British Isles

*Rhododendron ponticum* is a non-indigenous evergreen shrub. It was introduced into the British Isles around 1763 as a cultivated flowering plant and was widely planted in gardens, parks and estates (Figure 2a). It was also planted extensively in Victorian hunting estates, particularly in western coastal areas, under woodland canopies and on heathland areas to provide shelter for game species. It was later used as a rootstock for less hardy scions of *Rhododendron* species and cultivars, but the original rhododendron rootstock often produced its own shoots, which out-competed and then replaced the species or cultivar and reverted back to wild type rhododendron (Figure 2b).

During its cultivation and use as a rootstock, rhododendron was crossed with several other *Rhododendron* species to produce a range of rootstocks called the 'iron-clads' or Hardy Hybrids. These hybrid plants were further used in cultivation and plantings. The severely cold winter of 1879–80 killed many evergreen plants, including *Rhododendron* species, being planted at that time (Milne, 1997) and only some of the hybrids appear to have survived. These surviving hybrid bushes produced viable seed and enterprising gardeners collected their

**Figure 2**

Gardens with rhododendron specimens can occasionally contain *R. ponticum* (a), which can disperse seed into adjacent areas. *R. ponticum* rootstocks often produce vigorous shoots independent of the horticultural specimen grafted onto them, flowering and eventually overwhelming the specimen plant (b).

naturalised seedlings for distribution around the country. A further severe winter in 1894–1895 damaged and distorted some *Rhododendron* species, and the popularity of the genus dwindled and planting was reduced.

DNA analyses of naturalised 'wild type' rhododendron (Milne, 1997; Milne and Abbot, 2000), indicate that most of the naturalised bushes in the British Isles are derived from material originally introduced from the Iberian Peninsula. A smaller proportion has been shown to come from Turkish material. They also suggest that most of the naturalised material in the British Isles shows signs of introgression with other rhododendron species, particularly *R. maximum* and *R. catawbiense*. This is especially noteworthy for the eastern Scottish populations of rhododendron, which have had a high introgression rate with *R. catawbiense*, which may confer improved cold tolerance. The altered gene pool thus created would increase the competitiveness of rhododendron in Britain, everything else being equal (Milne, 1997).

## The need for control

Although the flowers are attractive, rhododendron has few other attributes that offset the negative impact it can have on a site that it invades (Figure 3). It is an aggressive coloniser

**Figure 3**

Dense rhododendron on a former clearfell site in west Argyll. This site has not been restocked, and will now require considerable expense before planting is possible.

that reduces the biodiversity value of a site; it obstructs the regeneration of woodlands and once established is difficult and costly to eradicate. As bushes mature and occupation of an invaded site increases, physical access can be reduced by the sheer density and size of the plants present, and the cost of some management operations (e.g. felling) can increase if the bushes require treatment first. Mature bushes also act as a prolific seed source for invasion of adjacent areas, and are a continued source of new plant material into areas successfully cleared.

## Rhododendron biology

*Rhododendron ponticum* is an evergreen shrub belonging to the Ericaceae family, which includes other species such as heather (*Calluna vulgaris*) and bilberry (*Vaccinium myrtillus*). Mature plants that are already established on suitable sites expand their area of occupation either by vegetative spread through stem layering or by seed dispersal and seedling establishment.

### Stem layering

The rate of spread by stem layering is relatively slow, as it requires the constant collapse and rooting of branches on the periphery of the bush. In broadleaved woodland the build-up of moist leaf litter provides an ideal medium for layering to occur. Where there is support from adjacent bushes or trees, the opportunity for branch collapse is reduced, and branch elongation rather than bush expansion occurs. In more open conditions, stem layering (Figure 4) ensures the clonal spread of particular bushes; the initial central root system may even degenerate and die but the newer vigorous ramets will ensure bush continuity on site. Because of its slow expansion, this type of spread often goes unnoticed, but can be an important mechanism for colonisation of sites that are not suitable for invasion by seedlings.

### Seed production

Rhododendrons produce monoecious, compact racemes of flowers annually in May–June (Figure 5). Flowering begins between the age of 12–20 years but may occur as early as 10 years on favourable sites (Figure 6). Regrowth from cut or flailed stumps of mature bushes may begin flowering within only two growing seasons after severance. A mature rhododendron bush growing in ideal conditions with space to expand can produce up to one

**Figure 4**

Stem layering is slow but it is an important cause of bush expansion.

**Figure 5**

Racemes of individual flowers are produced at the terminal end of shoots.

**Figure 6**

Flowering age of bushes at Mamore Estate in Scotland. Bushes were aged by destructive sampling and categorised as having flowers present or absent.

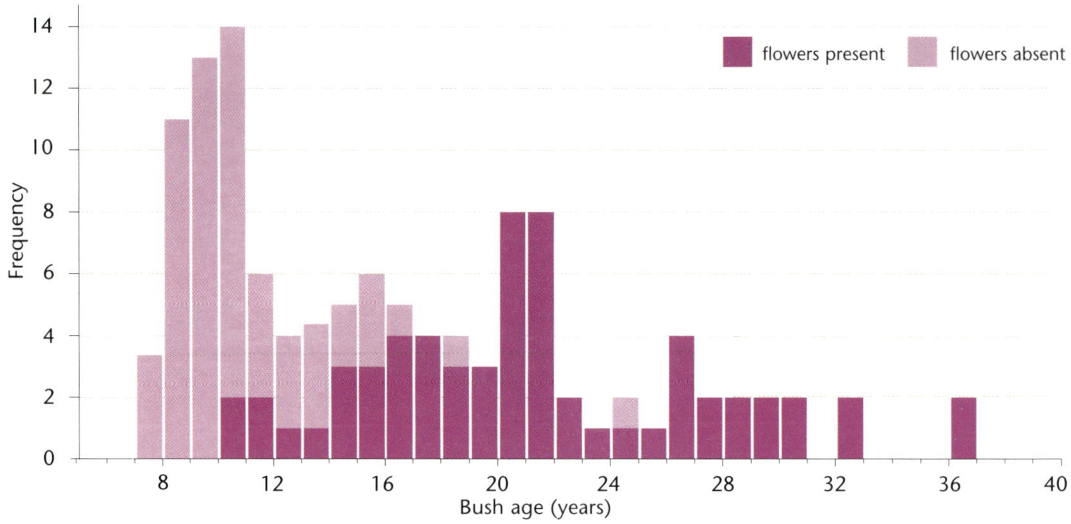


million seeds per year. Even a small plant can produce many thousands of seeds. If conditions are favourable for seed germination and seedling growth, and flowering plants are close by, invasion or reinvasion of a site can be rapid.

Rhododendron produce viable seeds annually, unless they are growing in unfavourable or unsuitable conditions. The combination of annual flowering and prolific seed production per bush leads to a high probability of seedlings reinvading sites from nearby seed sources.

## Seed dispersal

The seeds ripen in December and dispersal begins in February or March depending on weather. Rhododendron seeds are amongst the smallest and lightest of any plant species (0.063 mg), and are designed primarily for dispersal by wind (Figure 7). However, this appears to be limited to 5 m within woodland and up to 100 m in open flat ground on a lightly breezy day (Kohn *et al.*, personal communication). High winds and dispersal from high altitude seed sources have the capability to disperse seed further, especially if they get into upper air currents, although it is unlikely that many seeds will ever be carried more than 500 m from the source plant by wind (Figure 8).

**Figure 7**

(a) Flower petals have recently fallen from the inflorescence leaving the seed capsules visible. (b) The seed capsules then split and reflex open to enable seeds to be dispersed by wind.

**Figure 8**

Bush or seedling age and distance from the original parent seed source, which is in a walled garden, at Mamore Estate. Plant invasion onto adjacent heathland has travelled 1200 m in 50 years (= ~24 m $yr^{-1}$).

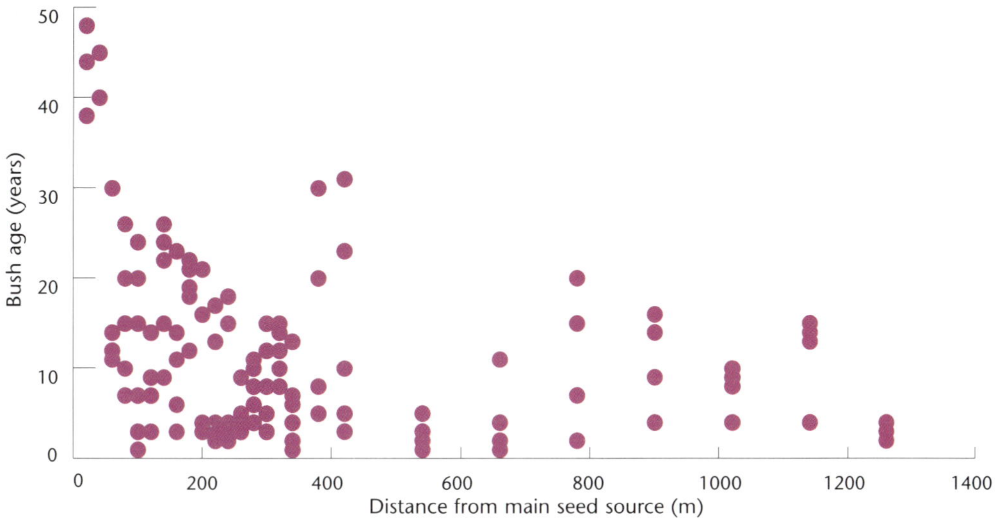


More recent evidence suggests that long distance dispersal of rhododendron seeds over several hundred metres can take place, but these events are likely to occur via a dispersal agent other than wind (Kohn *et al.*, personal communication), e.g. on animal hair, on the clothing of volunteer workers, or in mud on the tyres of a forest machine. Information on seed dispersal can be used to predict the future progression of invasion on to land under different land management regimes. It should also be possible to determine the width of a strip of forest needed to provide an effective barrier between an area clear of rhododendron and one that is infested.

## Seed germination

Once wetted, shed rhododendron seeds only remain viable for up to one year, but can germinate on a wide variety of substrates (Cross, 1975; 1981). They are reliant on light and constant humid conditions for germination (Colak, 1997; Cross, 1981).

## Germinant growth

Germinant establishment is thought to be restricted to thin layers of bryophytes covering bare ground (Figure 9), rock outcrops or tree stumps (Cross, 1975; 1981). If germination occurs on deep (>1 cm) bryophyte mats, establishment is unlikely as roots may fail to reach soil and germinants will die during a dry spell. The features of habitats most suitable for rhododendron establishment have been described qualitatively on a number of occasions (Cross, 1973, 1975; Rotherham, 1986, 1990; Rotherham and Read, 1988; Shaw, 1984; Thomson *et al.*, 1993).

**Figure 9**

Seedlings germinate well in the humid conditions provided by some low-growing bryophyte species.

## Seedling growth

Work carried out on the west coast of Scotland in 2003 showed that the probability of finding a seedling relates mainly to substrate type, substrate depth and the distance to the closest seed source (Stephenson *et al.*, 2005). The principal substrate for seedling establishment appears to be thin layers of moss or litter on dead tree material such as tree stumps.

Seedling growth is initially very slow (Figure 10) but becomes more rapid as time progresses (Figure 11). In sheltered conditions the plant's spreading habit can quickly lead to canopy closure and shading of the ground vegetation beneath. The combination of this shading and competition with rhododendron's shallow rooting system leads to a gradual loss of the associated ground flora, followed by a steady build up of organic matter and a mor humus.

**Figure 10**

Early rhododendron seedling growth from germination. (Adapted from Cross, 1981 – Table 2.)

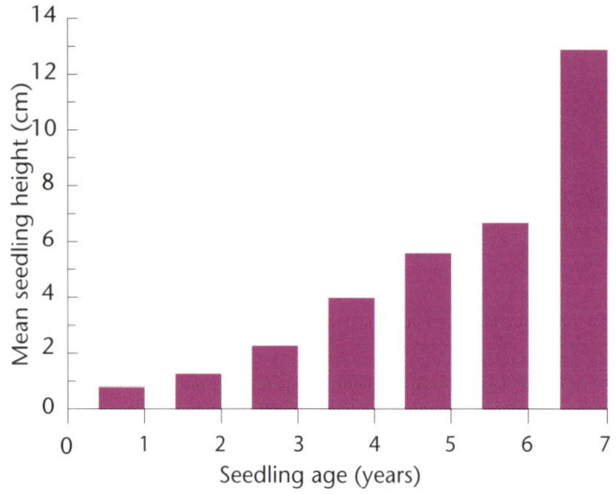


**Figure 11**

Bush height growth from germination to maturity for rhododendron bushes growing on a range of sites in Scotland.

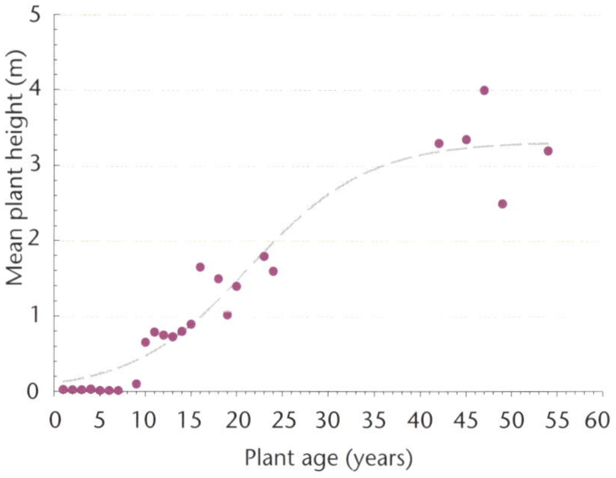


In open environments, plants develop into compact rounded bushes; this is especially noticeable in exposed areas (e.g. hill tops and coastal areas). Where support from neighbouring bushes or tree species exists (Figure 12), bush height growth exceeds lateral spread resulting in tall 'leggy' plants which can often reach 7–8 m in height in ideal conditions. Very mature bushes of this type can develop substantial stem diameters and become very difficult and expensive to manage by flailing or cutting and burning.

The biological longevity of rhododendron bushes in the UK is uncertain because very few have died due to natural causes, however there are live bushes forming a hedge in the National Trust for Scotland Gardens at Inverewe that are known to be 130 years old. Inverewe is noted for having a mild climate and good growing conditions, so this approximate age may not reflect the plant's longevity in more exposed areas. However, it does suggests that once rhododendron bushes are established, only direct control will remove them and return the site to a near natural or pre-invasion condition.

**Figure 12**

With support from surrounding trees or neighbouring bushes rhododendron can reach heights of 8 m or more. (a) A wall can also provide support as show here. (b) In this extreme case, support from an oak tree has allowed the rhododendron to reach the upper canopy.

# Managing and controlling rhododendron

It is important that a management plan is drawn up before control operations begin. All management plans should begin with a survey of the site and the condition of the bushes. Several sites may require treatment, but it may be impractical to treat them all in any one period. A priority must then be set on those that should be cleared immediately and those where treatment can be deferred until later. Once an order of priority has been decided, the appropriate methods of treatment should be chosen. This will include selecting the correct herbicide, method and timing of application. A site may contain different bush types depending on the conditions they are growing under, and colonisation of adjacent ground may be occurring at different rates. Each bush type is most efficiently controlled using different techniques so a combination of several control methods may be required for clearance operations to be fully effective. Finally, a monitoring schedule should be established to check the effectiveness of the control strategy and inform the revision of the plan. The seven stages to planning and implementing a control operation are outlined in Figure 13.

**Figure 13** The seven stages of a control operation.

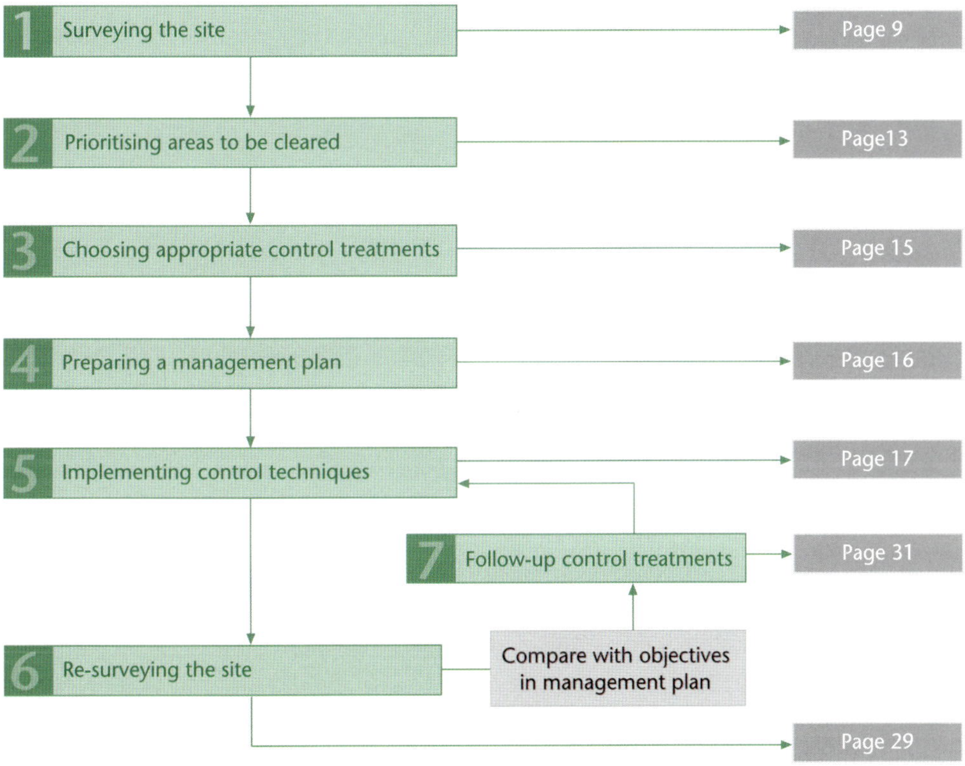


## Costs of control operations

Linking the survey data with clearance costs for each bush type will allow precise costs for the entire control area to be estimated. This costing exercise is vital as it can then be used to ensure sufficient funding is available for treatment of each of the bush types and areas for control over the life span of the management plan.

## Other considerations when planning control operations

Managers must also consider the social impacts of clearance operations. Many locations are visited annually by groups specifically to view rhododendron in flower. Informing stakeholder groups of intended operations and explaining the management and ecological reasons for clearance operations is essential if potential confrontations are to be avoided.

Good examples of areas where this has been achieved are the Beddgelert Community Group in North Wales and the Sunart Project in northwest Scotland. In both cases the local community understands and supports the removal of rhododendron bushes from large areas of wooded and non-wooded habitats.

## Success of control operations

It is useful to think of rhododendron control in terms of a simple predictive model:

$$\mathbf{I} = (\mathbf{S} \times \mathbf{R}) - \mathbf{K}$$

Where **I** = plant invasion of a site
**S** = seed source or vegetative expansion
**R** = rate of seedling recruitment
**K** = mortality, either natural or artificial (i.e. control operations)

The predictive model indicates that rhododendron plant invasion onto a site (**I**) will occur if there is a seed source (**S**) within dispersal distance of a site and this seed can germinate and grow (**R**), and the combination of these factors is greater than the rate of plant mortality (**K**), either natural or artificial.

If rhododendron bushes are growing on a site it follows that there must be a seed source and appropriate conditions for recruitment (Figure 14). Simply increasing the rate of mortality (i.e. bush control or clearance) without considering the seed source and/or suitability of the site will only be a temporary solution. Therefore to prevent reinvasion on a particular site, seed sources need to be removed, and/or seed germination prevented, or sustained mortality by control must be greater than the rate of seed dispersal and germination.

Complete eradication of all rhododendron from a site or catchment area will require careful planning if success is to be achieved. An effective management plan should run for at least seven years, identify the priority areas to be treated and detail the most effective control methods.

**Figure 14**

Expansion of rhododendron bushes onto open heathland from a forest plantation in west Argyll. Plantations often harbour scattered or individual bushes that can act as the seed source for expansion and invasion into receptive sites.

# Stage 1: Surveying the site

Most current rhododendron survey techniques are based on recording only the size of bushes and occasionally their distribution or density across the site. Although these surveys give adequate information for simple clearance operations, they fail to identify the most effective control operations available and do not allow managers to prioritise areas or decide whether the area should be treated at all.

The survey technique suggested in this Guide is based on the classification of bushes into suitable size classes for planning an optimal control strategy, by linking bush types with the most effective control methods and relating seed production to the order of treatment. Areas are further separated into those that are colonising rapidly, and those where bush development is slow.

## The steps of a site survey

- Map the boundary of each control site and the location and type of each rhododendron bush, or group of similar bushes, using the descriptions in Table 1.
- Use the decision chart in Figure 15 together with the associated classification table to categorise each control site into Site classes.
- Estimate the density of plants in each control site, and record other information such as machine access, presence of watercourses, location of areas for safe burning.

Figure 16 (page 12) illustrates a hypothetical example of a site survey map showing boundaries of control areas, site classes and bush densities.

**Table 1** Description of rhododendron bush types for clearance options.

| Bush type | Description |
|---|---|
| None present | Site clear of bushes. |
| Freshly cut stumps | Recently cut stumps that can be treated with herbicide within a 7-day period from cutting. |
| Old cut stumps | Stumps from previous control operations that may have buds but do not yet have live shoots suitable for treatment. |
| Stumps with regrowth | Shoot regrowth from previously cut stumps that is below 1.3 m height and not yet flowering. |
| Seedlings | Seedlings that have recently germinated and established. These tend to be shallow rooted and relatively young. |
| Small bushes | Bushes below 1.3 m in height with less than 3 m bush circumference. Access to entire circumference would permit all live foliage to be treated. |
| Medium height bushes or broad groups | Bushes or stump regrowth above 1.3 m in height, with no access to individual stems, or small bushes or groups with a circumference greater than 3 m. |
| Mature bushes | Bushes above 1.3 m height, with stems of >3 cm diameter that can be individually accessed. |

**Figure 15** Description of the site, site class, potential site development, and suggested priority and control options.

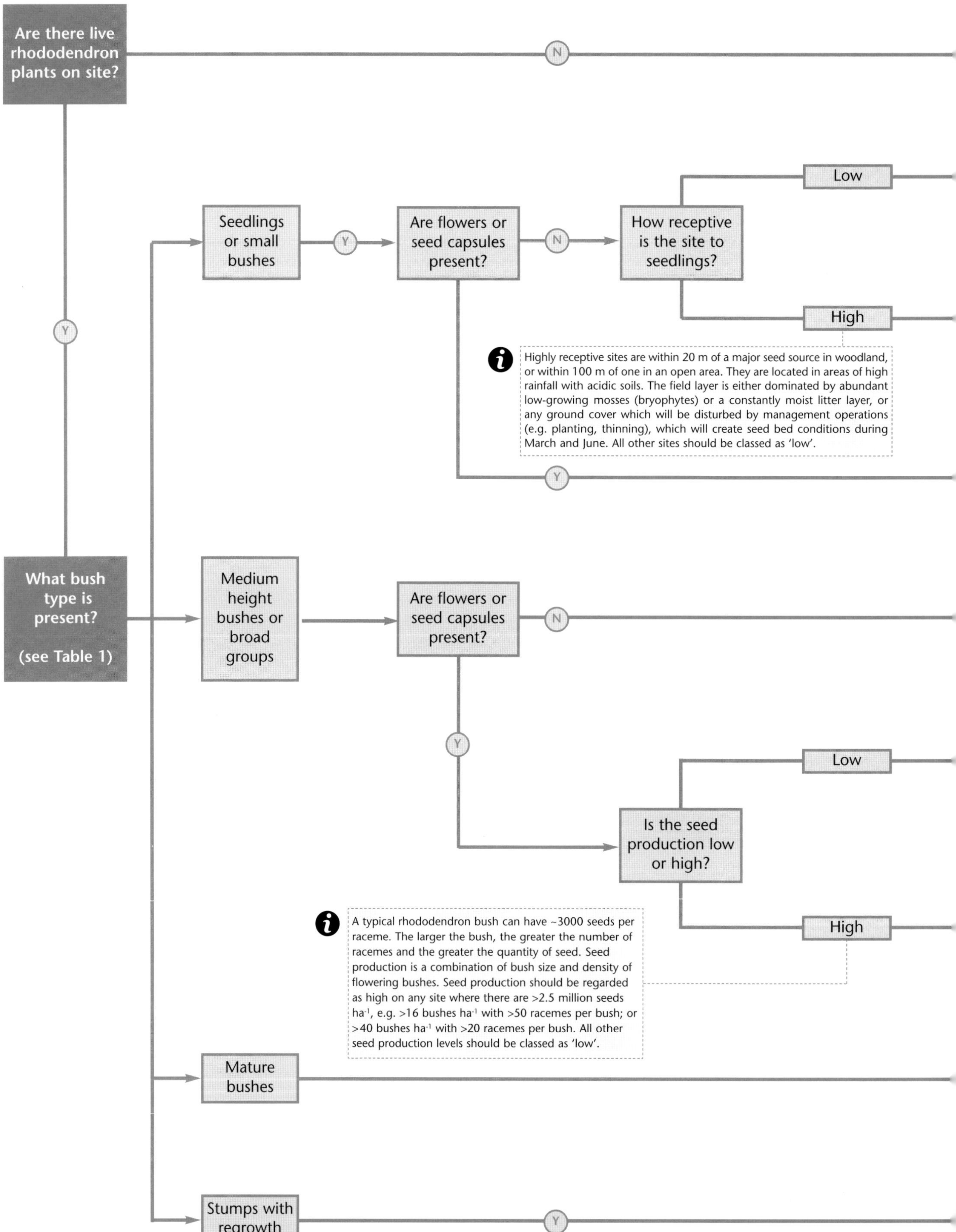

| Class | Detailed site description | Potential site development | Priority for control |
|---|---|---|---|
| **0** | Site clear of bushes. | Ensure sites remain clear of seedlings if adjacent to Site classes IV–VII by avoiding major disturbance events that will create favourable conditions for seedling establishment. | No control required. |
| **I** | Sites that are poorly receptive to seed influx, germination and establishment of bushes. Bush growth likely to be very slow (see Box 1) and age to flowering extended. Bush densities likely to be low, usually an expanding margin onto an unsuitable habitat for major rhododendron growth. | Unlikely to develop beyond Site class III. | Very low priority sites that are easily treated by foliar applications with herbicide. Small bushes and small seedlings can be easily pulled by hand. |
| **II** | Sites that are highly receptive to seed influx, good for germination and establishment of bushes. Probably recently disturbed sites, or sites with ground cover of thin layered bryophytes. Current bush growth rapid (Box 1), but still early in stage of bush development, >6 years before seed production and further expansion expected. | Will develop in time into Site class IV which then requires bush removal before herbicide control is possible. | Low-priority sites that are easy to treat but should not be left too long or they will be too large to treat without cutting down. Size currently within limits for overall foliar applications with herbicides, unless caught early likely to be too large to hand pull. |
| **III** | Sites that are acting as minor seed sources, dispersal and potential establishment of bushes onto adjacent sites very low. Bush growth slow (Box 1) and mainly vegetative, although a few flowers produced annually. Likely at limits for bush survival, either on excessively dry or wet soils, high pH or exposed sites. Seed production low. | Should not be left if very close to recently cleared receptive sites. Unlikely to develop into Site class V unless major site changes. | Medium-priority sites that are easy to treat, but should be left until higher priority sites cleared. Size within limits for overall applications. |
| **IV** | Sites that are highly receptive to seed influx and potentially new germination and establishment of bushes. Bush growth rapid with <4 years to flowering age likely (Box 1). Late in stage of bush development with limited time before seed production. | Will rapidly become Site class V. | Medium-priority sites with bushes that are too large to treat without cutting down. Size outwith limits for overall applications, some cutting, or flailing required before herbicides can be applied. |
| **V** | Sites that are acting as minor seed sources, dispersal and potential new establishment of bushes onto adjacent sites low. Bush growth rapid although flowering still only partial. Seed production low, but potential for increase in subsequently years. Likely to be under dense shade from a tree canopy, which if reduced or completely removed will result in rapid growth and onset of flowering. | Likely to develop rapidly into Site class VI or VII. Should not be left if very close to recently cleared receptive sites. | High priority sites that are not easy to treat, but can be left while still producing few seeds until higher priority Site classes VI and VII cleared. Size outwith limits for overall applications, some cutting or flailing required before herbicides can be applied. |
| **VI** | Sites occupied by mature bushes that are acting as a major seed source. Bush growth rapid, with a high stem density and/or complete canopy cover. Seed dispersal and potential new establishment of bushes onto adjacent sites very high. Seed production high, 750K–1.5 million seeds $bush^{-1}$ $year^{-1}$. | Expected to develop rapidly into Site class VII. Should not be left if very close to recently cleared receptive sites. | Second highest priority sites that are not easy to treat but should be cleared first along with Site class VII. Access to stems not always possible to allow stem treatment treatment. Size outwith limits for overall applications, some cutting or flailing required before herbicide can be applied. |
| **VII** | Very mature sites that are acting as major seed sources, dispersal and potential new establishment of bushes onto adjacent sites very high. Bush growth into woody stage, probably associated with an overstorey of trees. Seed production high, 750K–1.5 million seeds $bush^{-1}$ $year^{-1}$. | Should not be left if very close to recently cleared receptive sites. | Highest priority sites that are not easy to treat unless stem treated, should be cleared initially along with any Site class VI. Size outwith limits for overall applications, some cutting or flailing required before herbicide can be applied. Stem treatment application of herbicide possible on these bushes, avoiding having to cut down bushes before control. |
| **VIII** | Foliar regrowth from stumps originally classed as III, IV, V, VI or VII. | Should not be allowed to develop to a height >1.3 m or to begin flowering. Flowering will begin when regrowth is 2 or 3 years old. Shoot elongation likely to be rapid: 20–40 cm $yr^{-1}$ (Box 1). | High priority. Treat using a foliar regrowth application before shoots reach 1.3 m height. Once shoot regrowth is >1.3 m its size will be outwith limits for overall applications, some cutting or flailing would then be required before herbicide can be applied. |

**Box 1** Assessing the speed of seedling growth (based on Erfmeier and Bruelheide, 2004).

Fast growing seedlings have a mean shoot extension rate greater than 15 cm $yr^{-1}$.

To calculate mean extension rate, measure the previous season's growth on the highest shoot with a terminal vegetative bud or, if sampling during the growing season, measure the growth between the beginning of this year's current growth and the previous year's scar. Sample a minimum of 20 plants to calculate the mean.

**Figure 16**

A hypothetical example of a site survey map showing boundaries of control areas, site classes and bush densities (% cover).

Map legend

- A Site class VII: mature bushes
  Acting as major seed source.
  Access possible to stem treat only 70% of area, remaining 30% of area must be manually cut and burnt.
  Material to be dragged and burnt on adjacent clearfell.
- B Site class V: medium-height bushes
  Low seed production. Machine access possible.
- C Site class II: seedlings
  Rapid invasion onto disturbed ground. Rapid plant growth.
- D Site class I: seedlings
  Slow plant invasion into site. Slow plant growth.
- E Site class N/A
  Recent restoration site with a clearfelled area suitable for burning material from Site class VII area.

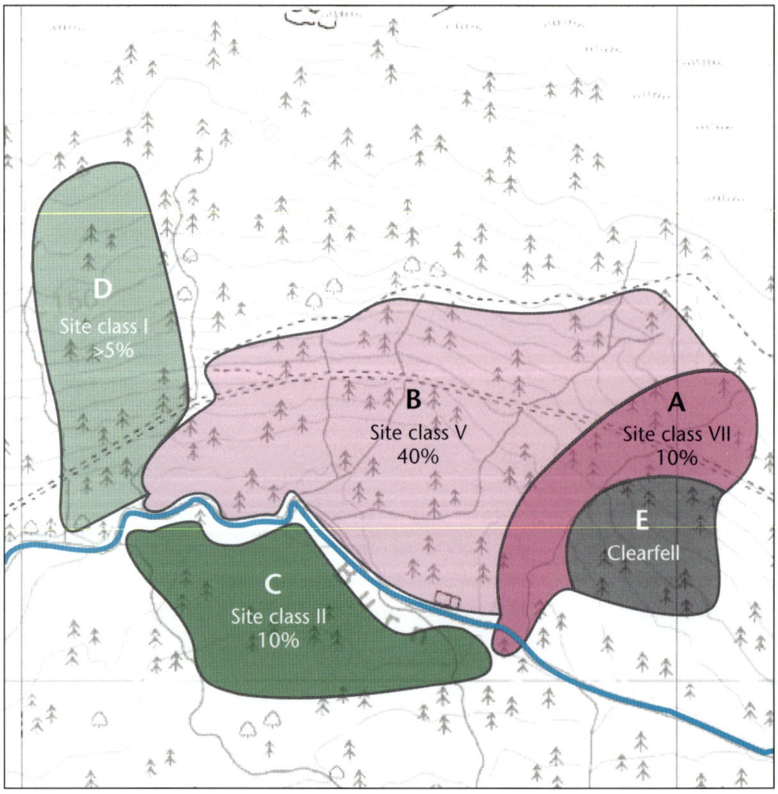

# Stage 2: Prioritising areas to be cleared

The recommended order of priority in which to treat bushes is given below; guidance on these priorities and exceptions to the order are also provided.

## Treat newly cut stumps promptly (Site class VIII)

If bushes have been recently cut or have untreated regrowth that is <1.3 m, treat with herbicide as a priority before they become too large for an overall foliar treatment and either need to be cut again or begin flowering and distribute seed (Figure 17).

**Figure 17**

(a) Regrowth on a stump cut 5 months previously: too small to treat with herbicide. (b) Regrowth on stumps 18 months after flailing; this is the perfect time for foliar spray treatment with herbicide.

## Treat major seed sources next (Site classes VII and VI)

The largest and oldest bushes will be producing the most seed (Figure 18) and pose the greatest threat to the long-term eradication of bushes within a single control area. Tackling the mature seed-producing areas before minor seed sources prevents the distribution of seed onto receptive sites, reducing the requirement for future interventions on an ever-increasing area. This is particularly important if adjacent areas are either to be managed or disturbed as this will make them receptive to seed germination and growth of new rhododendron plants.

**Figure 18**

Prolific flowering on this individual bush indicates the potential for high seed production; each raceme here will produce in excess of 3000 seeds.

## Treat minor seed sources soon after major ones (Site class V)

Once the major seed sources have been treated, minor ones will become the primary sources of seeds and should be tackled swiftly. There are however two situations when Site class V should be given the same priority as VII or VI; either they are adjacent to a receptive site or the bushes are stressed, e.g. under a dense canopy as in Figure 19. Because stressed plants may be easier to kill than those growing in ideal conditions, treat them before any management operations alter their site conditions and strengthen their growth, making them more resilient to control treatments.

## Treat fast growing bushes before slow growing bushes (Site classes IV and III)

Fast growing bushes are likely to reach flowering age sooner and will put on more growth, and hence produce more seeds, than bushes in less ideal locations. If small bushes are growing especially quickly it is worth treating them earlier than those in Site class V if either they are adjacent to a receptive site or a delay will significantly increase treatment costs.

## Treat highly receptive sites before poorly receptive ones (Site classes II and I)

Once all principal seed sources have been removed, smaller non-flowering bushes (Figure 20) can be targeted for treatment. Since rhododendron do not form a persistent seedbank, only seedlings that have recently germinated will recolonise the site. As these seedlings can be very small (<10 cm tall at age six years; see Figure 10) they are difficult to locate and treat. These are best left to develop into larger plants that can then be pulled or sprayed.

**Figure 19**

In non-ideal conditions, flower production is lower and bushes are weaker and likely to be more susceptible to herbicide treatments.

**Figure 20**

Small isolated bushes growing in unfavourable conditions are a low priority in an eradication programme.

Use the detailed information in Figure 15 and its associated classification table to identify the order in which areas should be tackled. In general, the Site classes with the highest scores are of higher priority. These tend to be the major infestation or seed source in an area. The lower priority bushes will normally be located on the periphery of these infestations either as an expanding front or as small and isolated bushes growing in less ideal conditions. If the management objective is the eradication of all rhododendron bushes in an area, it is essential to follow the order of priority given even though it is easier and cheaper in the short term to control small bushes and seedlings first. Following an alternative priority order is likely to result in continued expenditure in clearing bushes from an ever-increasing area.

# Stage 3: Choosing appropriate control treatments

The safest and most efficient methods of controlling rhododendron depend on the size, life stage and accessibility of the target bush. In general stem treatment is the most effective and efficient method of killing large rhododendron bushes. Where there is no access to treat the stems, an overall foliar spray is the next option. Small bushes are easily treated, but foliar sprays cannot be applied safely to larger bushes, which must therefore be reduced in size before a safe application of herbicide can be given. There are two main ways to reduce bush size: using a mechanical flail or by manual cutting.

In some circumstances the only means of approach to bushes to apply any treatment is by roped access, e.g. in ravines or on cliff faces. Teams of qualified operators now offer this specialised service. The most successful operators kill bushes *in situ* with one visit: by stem treatment using cordless drills rather than other techniques that require several visits.

Use the information collected during the initial site survey to identify the most appropriate control technique. Figure 21 identifies the recommended control technique for each of the bush types described in Table 1 (page 9). Then the area to be cleared can be divided into sections, each section representing a bush type which will be controlled in one standard technique.

**Figure 21** Recommended control methods for bushes of a specific type. See also Table 1 (page 9).

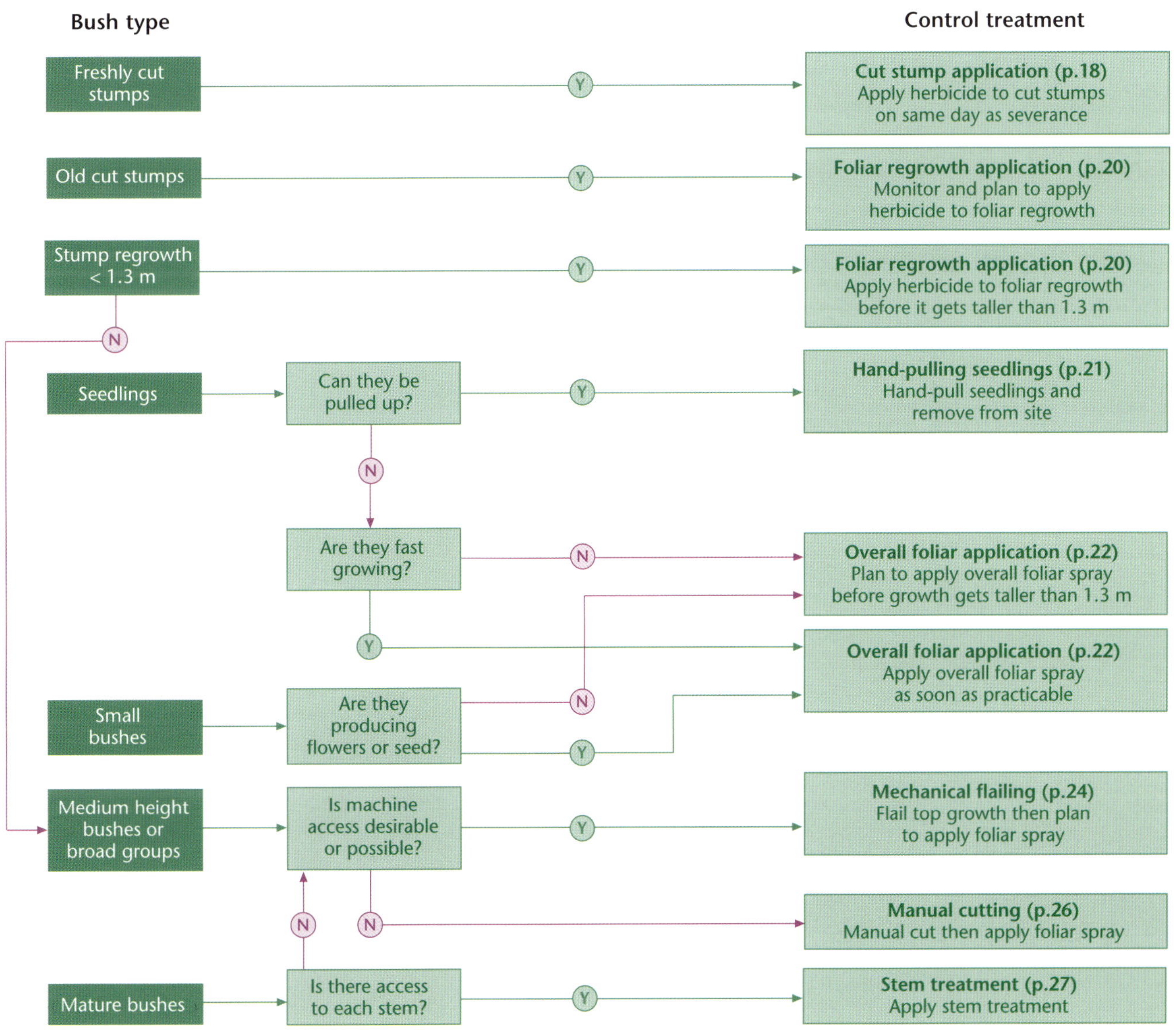

# Stage 4: Preparing a management plan

After surveying the potential areas to be controlled, decide which are priority areas and which control techniques are to be used, and then write a management plan. The plan should record the area to be controlled, the site classes, control methods and their treatment order (Table 2). The management plan must take account of the following factors.

## Machine access

Flails cannot be used where there is no access from the road onto the site, or where slopes are beyond the acceptable limits for the base machine.

## Prevailing wind direction

Bush clearance operations should work with the direction of the prevailing wind to prevent seed distribution from an existing rhododendron canopy being blown onto a recently cleared site.

## Constraining factors

If the intention is to burn material after it is cut there must be sufficient suitable areas for locating the fires. See Manual cutting (page 26) for details on burning restrictions.

## Time of year of operations

The action of cutting or flailing bushes following a spell of dry weather in February to May will generally aid the distribution of seed from open seed capsules. In operations that include the removal of bush top growth during this period of seed dispersal, caution may be needed to prevent spreading seed across larger areas by inadvertently transporting seed-laden branches around the site. Before such operations are undertaken, the bushes should be checked for open seed capsules and, if present, for seed. Managers must be in no doubt that the probability of seed dispersal is offset by the advantages of removing the bushes at this time. Otherwise operations should be timed for outwith the seed dispersal period.

The objective of all control operations should be to visit the target site only once, or as few times as possible, to achieve bush control. Repeated applications of herbicide to persistent bushes should be avoided. When continual applications have become necessary the control technique must be critically appraised; perhaps the correct herbicide or application technique is not being employed or there is a more effective alternative method.

**Table 2** A summary of the management plan in tabular form based on the example in Figure 16.

| Area | Site class | Comments | Priority | Control | Year |
|---|---|---|---|---|---|
| A | Class VII:<br>mature bushes | Acting as major seed source. Access possible to stem treat only 70% of area, remaining 30% of area must be manually cut and burnt. Material to be dragged and burnt on adjacent clearfell. | 1 | • Stem treat 70% area.<br>• Cut and burn 30% area.<br>• Cut stump treat 30% area.<br>• Foliar regrowth application live stumps. | 1<br>1<br>1<br>3 |
| B | Class V:<br>medium-height bushes | Low seed production.<br>Machine access possible. | 2 | • Machine flail.<br>• Foliar regrowth application any live stumps. | 2<br>4 |
| C | Class II:<br>seedlings | Rapid invasion onto disturbed ground.<br>Rapid plant growth. | 3 | • Overall foliar application. | 3 |
| D | Class I:<br>seedlings | Slow plant invasion into site.<br>Slow plant growth. | 4 | • Hand-pull seedlings with volunteers. | 4 |
| E | N/A | Recent restoration site with a clearfelled area suitable for burning material from A. | – | – | – |

# 5: Implementing control techniques

Figure 21 on page 15 should be used to select the appropriate control technique for a particular bush type. Each control technique is described in detail in this section, including approved herbicides and rates of application, equipment and additional considerations necessary to achieve complete control.

Table 3 provides information on herbicides that have been evaluated in trials by Forest Research, and gives details of hazard rating and selectivity restrictions on use. It is essential to check product labels for the most up-to-date information on current approval status. For safety information on using herbicides consult the Defra publication *Pesticides: code of practice for using plant protection products*, AFAG leaflet 202 *Application of pesticides by hand-held equipment* and the HSE leaflet *Pesticides: use them safely*. Always read the label before applying any herbicides.

**Table 3** Herbicides and adjuvant tested in Forest Research trials.

| Herbicide | Hazard classification for products | Selectivity |
|---|---|---|
| 2,4-D/dicamba/triclopyr | Irritant to eyes and skin<br>Harmful if swallowed<br>Harmful to aquatic life | Annuals, perennials and woody weeds |
| Adjuvant (High Trees Mixture B)[1] | Irritant to eyes and skin<br>Harmful if swallowed<br>Harmful to fish | |
| Ammonium sulphamate[2] | None<br>Harmful to aquatic life | Non-selective |
| Glyphosate | Roundup ProBiactive and Envision – none<br>For other products, refer to FC Practice Guide *Reducing pesticide use in forestry* and product labels | Non-selective |
| Triclopyr | Irritant to eyes and skin<br>Harmful if swallowed or in contact with skin<br>Harmful to aquatic life | Perennials and woody weeds |

[1]When near water do not use Mixture B – apply 2.5% solution of glyphosate instead.
[2]Indications are this product is unlikely to be given Annex 1 listing by Pesticides Safety Directorate (PSD), and may no longer be available after 2006.

The Forestry Commission Practice Guide *Reducing pesticide use in forestry* (Figure 22; Willoughby *et al.*, 2004) gives detailed guidance on the environmental risk of different herbicides, and how to select herbicides to lessen risk. The Forestry Commission pesticides resource site (www.forestry.gov.uk/pesticides) has links to various on-line product labels and other sources of advice. Forestry Commission Field Book 8 *The use of herbicides in the forest* (Willoughby and Dewar, 1995) also contains useful information on safe and effective herbicide use.

**Figure 22**

Forestry Commission Practice Guide: *Reducing pesticide use in forestry.*

If herbicides are to be used in any water catchment area where there is a risk of water contamination, it is essential to inform the relevant regulator (the Environment Agency in England and Wales, the Scottish Environmental Protection Agency (SEPA) in Scotland and the Department of Agriculture and Rural Development in Northern Ireland) before the start of operations. It is recommended that, where applications are made within 10 m of a permanent watercourse or 20 m of standing water, only a glyphosate-based product with no hazard rating for aquatic life (if in doubt check the product label) should be used. The method of application will have an impact on risk of water contamination – correctly managed stem treatment is likely to have less risk of run-off than other methods.

## Cut stump application

This technique involves applying an approved herbicide to the surface of a stump on the same day of cutting (Figure 23). Drilling a reservoir on the stump surface to contain the herbicide has been shown to be a highly effective technique (Figure 24).

**Figure 23**

A close-up of an old cut stump surface. Only freshly severed stumps should be treated with herbicide.

**Figure 24**

These stumps were drilled before treatment to create reservoirs for the herbicide.

### Site factors

Stumps are often buried by earlier operations and can easily be missed or double treated during a cut stump application. To reduce these mistakes a suitable dye can be added to show where operators have applied herbicide; Dysol Turquoise ANX 50 at a rate of 2% of final spray volume (supplied by Albion Colours) can be added but a slight reduction in glyphosate efficacy may result.

### Equipment

Herbicide can be applied using:

- a knapsack sprayer at low pressure fitted with a solid cone nozzle;
- a forestry spot gun with solid stream nozzle;
- a paint brush;
- a clearing saw with suitable attachment;
- a watering can.

Operators must wear the correct protective safety clothing for mixing and handling herbicides during application.

In most situations use a knapsack sprayer for applying the herbicide solution. A spot gun should be used when only a few stumps are to be treated. The advantage of the paintbrush is its high degree of precision compared with any system using pressure, and it may be particularly suitable for areas adjacent to water, although stump saturation is difficult to achieve.

## Approved herbicides and application rate

Applications should saturate the stump cut surface and outer bark area, but should not create run-off into the surrounding soil. Avoid application to adjacent plant foliage or trees as damage is likely to occur at the approved rates. Ammonium sulphamate is absorbed by plant roots and should only be used as a pre-plant treatment. Dry crystals of ammonium sulphamate can be applied to holes drilled into target stumps at the rate of 6 g per 10 cm of stump diameter. Table 4 provides a list of approved herbicides and their application rates.

## Time of application

To be fully effective herbicides must be applied on the same day as stem cutting, in frost-free and rain-free conditions and when rain is not forecast for at least six hours. Applications immediately after cutting have a much greater chance of success. Delayed applications will be less effective, particularly with glyphosate.

**Table 4** Approved herbicides for cut stump applications.

| Herbicide | Product rate | Optimal time of year |
|---|---|---|
| Ammonium sulphamate<br>e.g. Amcide | 40% solution in water<br>6 g of crystals per 10 cm stump diameter | June–September |
| Glyphosate (360 g $l^{-1}$)<br>e.g. Roundup ProBiactive | 20% solution in water | October–February |
| Triclopyr (480 g $l^{-1}$)<br>e.g. Timbrel | 8% solution in water | October–March |

For safety information on using herbicides consult the Defra publication *Pesticides: code of practice for using plant protection products*; AFAG leaflet 202 *Application of pesticides by hand-held equipment*; and the HSE leaflet *Pesticides: use them safely.* Always read the label before applying any herbicides.

## Additional considerations

- Research experiments have shown that cut stump applications are rarely 100% effective and regrowth of shoot material should be expected on some stumps.

- Co-ordinate work so that fresh cut stumps are treated immediately after cutting, and at least on the day of cutting. If insufficient stumps are cut each day to warrant herbicide application, then consider re-cutting the stumps to create a fresh surface to treat. Although this increases the cost of the clearance operation, the herbicide application is likely to be more effective and fewer stumps will regrow, reducing the cost of follow-up foliar applications.

- Fresh herbicide solution should be used for each day's application.

- Rhododendron stems or branches that are still attached to the stump after cutting and those that have layered will not be killed by the application of a cut stump treatment. These stems or branches will need to be targeted with an overall foliar application. Stumps must be cut as low as possible to the ground, or at least below the last live branch to avoid leaving live branches attached.

## Foliar regrowth application

Foliar regrowth applications are similar to overall foliar applications, except that here the target material tends to be smaller and younger and has regrown from cut or flailed stumps. New shoot growth should be targeted when it has had at least one full growing season to develop (Figure 25) but must be treated before it reaches a height of 1.3 m – the safe height for overall spraying using hand-held applicators without undue risk to the operator. Regrowth is likely to begin flowering in the third season after initial severance, and must be treated before it reaches this stage of development (Figure 26).

**Figure 25**

Stump with regrowth at the ideal stage of development for a foliar regrowth application of herbicide.

**Figure 26**

Regrowth from treated (right) and untreated (left) stumps 24 months after flailing and cut-stump application of herbicide.

### Site factors

Foliar regrowth applications can be made on any site where foot access to the target bushes is possible.

### Equipment

Applications to foliar regrowth can be made with the following equipment:

- knapsack sprayer at low pressure fitted with a flood jet or solid cone nozzle;
- forestry spot gun with solid stream nozzle (small seedlings only).

Operators must wear the correct protective safety clothing for mixing and handling herbicides during application.

### Approved herbicides and application rate

Herbicide solution is applied to all live foliage as a directed spray to the entire bush until the instant just before run-off, i.e. foliar wetness. If using a knapsack sprayer, apply herbicide solution at low pressure and medium–high volume (500–750 l $ha^{-1}$). See Table 5 for further information.

The addition of the adjuvant High Trees Mixture B will increase the rate and speed of herbicide uptake and its use is recommended particularly with glyphosate.

**Table 5** Approved herbicides for foliar regrowth applications of herbicide.

| Herbicide | Product rate | Optimal time of year |
|---|---|---|
| 2,4-D/dicamba/triclopyr (200:85:65 g $l^{-1}$) | 7.5% solution in water (maximum of 5 l $ha^{-1}$) | July–September |
| Ammonium sulphamate e.g. Amcide | 40% solution in water | May–June |
| Glyphosate (360 g $l^{-1}$) e.g. Roundup ProBiactive | 2% solution in water plus 2% adjuvant* (High Trees Mixture B) (maximum of 10 l $ha^{-1}$) | March–October |
| Triclopyr (480 g $l^{-1}$) e.g. Timbrel | 2.5% solution in water (maximum of 8 l $ha^{-1}$) | June–September |

*When near water do not use adjuvant – apply 2.5% solution of glyphosate instead.
For safety information on using herbicides consult the Defra publication *Pesticides: code of practice for using plant protection products*; AFAG leaflet 202 *Application of pesticides by hand-held equipment*; and the HSE leaflet *Pesticides: use them safely*.
Always read the label before applying any herbicides.

### Time of application

Effective treatments should be made in frost-free and rain-free conditions and when rain is not forecast for at least 12 hours. However, manufacturers advise that triclopyr is rain-fast within two hours from application, therefore in conditions where rain is likely to fall within six hours of application, triclopyr may be more effective.

### Additional considerations

- The herbicides approved for foliar applications on rhododendron are not readily translocated laterally through the target bush. Experiments have shown that full bush recovery from partial foliar treatments may only take three to four years.

- Because of the high solution application rates, this technique is reliant upon an easily accessible supply of suitable clean water close to the treatment site.

- Applying herbicide solution at a greater than approved product rate will not increase the effectiveness of the application treatment.

- Young leaves on new shoots have thinner waxy cuticles than older material and allow faster and more complete penetration of herbicide.

- Foliar regrowth applications can target any small seedlings as well as the stump regrowth.

## Hand-pulling seedlings

Hand-pulling is a simple procedure – operators walk in line through the area to be cleared and hand-pull all live seedlings by gripping the base of the stem and pulling at an oblique angle.

Control of small or young seedlings can be achieved without using herbicides, but only if the area involved is small, the density of seedlings to pull is low or sufficient numbers of operators are available.

Larger seedlings or small bushes can also be pulled by hand, but winches or hand-held equipment, such as mattocks, are required.

### Site factors

This technique works best in loose or friable soils, especially when they are moist or wet, or when seedlings are rooted into leaf litter.

### Equipment

Hand-pulling requires:

- gloves;
- planting bag or similar;
- forestry mattocks, picks or similar hand-held tools that act on the lever principle;
- hand winches.

### Approved herbicides and application rate

None.

### Time of operation

Any time of year when enough staff or volunteers can be gathered together.

### Additional considerations

- Can be carried out by untrained staff and volunteers.

- There will be some soil disturbance. The extent will depend upon the soil type and size of root system on seedlings being pulled. Disturbed soil can act as a suitable seed bed for seedling germination, especially during the period of seed dispersal, i.e. February–May.

- A bag must be used to collect pulled seedlings for removal from site, or seedlings must be hung upside down on the branch of adjacent trees; if seedlings are left on top of the soil they can re-root in the humid/wet conditions where rhododendron is typically found.

- This is a useful operation on small areas that need to be kept seedling free. Seedlings must be allowed to grow to a suitable size before hand-pulling is considered, however, as searching for very small seedlings is ineffective and will rapidly demoralise volunteers.

## Overall foliar application

Overall foliar applications are similar to foliar regrowth applications. In this case, however, the target foliage tends to be slightly taller, though still below 1.3 m – the safe height for overall spraying using hand-held applicators without undue risk to the operator. Access must be available to the entire circumference of the bush (Figure 27). Approved herbicides are not translocated through rhododendron bushes so it is important that all live leaves are treated; incomplete applications of herbicide will result in only partial control of the target bush.

### Site factors

Overall foliar applications can be made on any site where pedestrian access to the target bushes is possible.

**Figure 27**

Foliar applications with glyphosate can kill large bushes if all live foliage can be safely treated.

## Equipment

Overall foliar applications can be made with the following equipment:

- knapsack sprayer at low pressure fitted with a flood jet or solid cone nozzle;
- forestry spot gun with solid stream nozzle (small seedlings only);
- watering can (ammonium sulphamate).

Operators must wear the correct protective safety clothing for mixing and handling herbicides during application.

## Approved herbicides and application rate

Herbicide solution is applied as a directed spray to the entire bush until the instant just before run-off, i.e. foliar wetness. When using a knapsack sprayer, apply herbicide solution at low pressure and medium–high volume (500–750 l $ha^{-1}$). The addition of High Trees Mixture B will increase the rate and speed of herbicide uptake and is recommended with glyphosate.

## Time of application

Effective treatments should be made in frost-free and rain-free conditions and when rain is not forecast for at least 12 hours. However, manufacturers advise that triclopyr is rain-fast within two hours from application, therefore in conditions where rain is likely to fall within six hours of application this product may be more effective.

## Additional considerations

- The herbicides approved for foliar applications on rhododendron are not readily translocated laterally through the target bush. Experiments have shown that full bush recovery from partial foliar treatments may only take three to four years.

- Applying herbicide solution at a greater than approved product rate will not increase the effectiveness of the application treatment.

- Because of the high solution application rates, this technique is reliant upon an easily accessible supply of suitable clean water close to the treatment site.

## Mechanical flailing

Woody rhododendron material is flailed or mulched on site using hydraulically powered flail heads mounted on tracked excavators (Figure 28). Flail heads are mounted onto either horizontal or vertical shafts. Horizontal shaft heads have recessed teeth, or hammers, on a shaft which shatter plant material (Figure 29). Vertical shaft heads have large, flat spinning discs with radially mounted teeth which work from the top of the bush downward, grinding material as they go.

**Figure 28**

A front-mounted flail working in a mixed woodland.

**Figure 29**

Front-mounted horizontal shaft flail with tungsten tipped teeth. A front door to the flail unit is open but when closed retains material in close proximity to the flail head for fine mulching.

The flailing operation leaves stumps available for immediate treatment by cut stump treatment (page 18) or for later treatment by foliar regrowth spray (page 20). Flailing on its own is insufficient to give complete bush control (Figure 30) and herbicide applications are essential for complete eradication.

### Site factors

Machine access onto the site from the roadside is required. Steep or boulder strewn slopes and unstable ground conditions are constraints to using standard JCB base units, although walking excavators fitted with flail heads are available (Figure 31). Horizontal shaft flail heads are highly manoeuvrable and not constrained by the presence of trees on site, however, machine access between trees will have to be considered.

**Figure 30**

An experiment site in Argyll. (a) A flailed area is shown in the foreground; the bushes in the background are untouched. (b) The same site 18 months later showing significant shoot regrowth from 90% of stumps.

## Equipment

- Hydraulically powered mechanical flails with either horizontal or vertical shaft heads.

## Additional considerations

- Some flail heads create a tangle of long shoot and stem pieces (Figure 32), making walking on the site difficult and creating a hazard to operators applying herbicide or to visitors in areas of public access.
- Stumps can also become buried by thick layers of mulch and can be easily missed in follow-up herbicide applications to cut stumps.
- Rather than applying a cut stump treatment only on those stumps that are visible, consider waiting for the stumps to regrow and then apply herbicide to the foliar regrowth.
- Mulch material will provide some vegetation suppression and may act as a barrier to seed germination of rhododendron and other plant species.
- Stems that become partially buried, or left on a damp substrate, may produce adventitious roots and form a new plant.

**Figure 31**

A Kaiser walking excavator fitted with a hydraulic flail head working on a slope too steep for tracked excavators.

**Figure 32**

Flailed stems showing the difficult site conditions in which operators must walk if herbicide applications are to be made to cut stumps.

## Manual cutting

Top woody growth of rhododendron is manually removed to leave cut stumps with no remaining live branches or shoots (Figure 33). Operators with chainsaws or clearing saws are usually employed to do this. The resulting cut woody material can be removed to a safe area for burning or chipped on site. Chips can be left on site or bagged and removed.

### Site factors

Burning is normally conducted in clearings, or in open areas without overhead trees. Freshly cut rhododendron material is difficult to ignite and benefits from being allowed to dry first. However, this can cause problems if there is insufficient space.

**Figure 33**

Volunteers at Callander Wood removing woody top growth of rhododendron for later burning.

### Equipment

Manual cutting requires:

- chainsaws or clearing saws;
- bow saws and loppers.

### Additional considerations

- Volunteers can help to cut rhododendron areas by hand, but large numbers of well-motivated volunteers are necessary to achieve effective control in large areas.

- Burning rhododendron waste is covered by the Waste Management Licensing Regulation (1994). The regulator (the Environment Agency in England and Wales, SEPA in Scotland and the Department of Agriculture and Rural Development in Northern Ireland) must be notified in advance and the amount burned in any 24 hour period must be less than 10 tonnes.

- Care is needed when positioning fires. Do not place them under an existing tree canopy or power or telephone lines. They should not be positioned close to houses, or within 50 m of a main highway, where the pungent smoke will be a nuisance.

- A build up of chipped rhododendron material can act as a mulch and prevent desirable vegetation species growing on the cleared site so, where possible, avoid this. Alternatively a mulch of chipped material can be used to reduce the potential encroachment of rhododendron or other weed species.

- The taller the remaining stump, the easier it will be to find it again for herbicide application (Figure 34). However this must be offset against the chance of leaving live branches attached below the height of severance.

- Stems that become partially buried, or left on a damp substrate, may produce adventitious roots and form a new plant.

**Figure 34**

Once cut, rhododendron stems must be removed from the site to enable access to stumps for follow-up herbicide applications.

## Stem treatment

Stem treatment is a relatively new technique that has been developed for controlling bushes where each stem is >3 cm diameter and can be individually accessed (Figure 35). The technique has been tested and found to be effective on the west coast of Scotland and North Wales. Some aspects of the technique, e.g. time of year for application, are still being investigated and the results presented here are based on limited evidence.

On small stems a hand axe or heavy-bladed implement should be used to create a wound into which the herbicide is applied. A cordless drill and bit can be used to drill a reservoir into larger stems into which the herbicide solution can be placed. A more detailed step-by-step guide is given in Appendix 1 (page 36).

## Site factors

This technique is particularly useful in locations where roped access is required to reach bushes. Teams of qualified tree arboriculturists and tree climbers now use stem treatment as the main control techniques for individual bush treatment in inaccessible locations such as cliff faces or ravines. Treated bushes are left *in situ*.

**Figure 35**

Large rhododendron bushes with access to individual stems are ideal for treatment by stem injection.

Stem treatment requires:

## Equipment

- hand-held cordless drill with sufficient rechargeable batteries for a full day's operation;
- hand-held axe or billhook;
- forestry spot gun.

Operators must wear the correct protective safety clothing for mixing and handling herbicides during application.

## Approved herbicide and application rate

A 25% solution of glyphosate is recommended. No additives are required.

## Time of application

Applications during March, April and October have been successful in giving complete control of target bushes. In all months applications are expected to give good control, but this has not been fully assessed. Winter applications are likely to take longer to have an effect than those during the growing season.

## Additional considerations

- One of the main advantages of this technique is the controlled application of herbicide to target bushes, and the reduction in damage to other adjacent plant species.

- Treatments early in the growing season prevent flowering, and hence reduce seed dispersal and the potential for colonisation of cleared sites at a later date.

- Dead bushes can be left on site to rot, or can be removed using manual cutting or flailing techniques at a later date. Dead material is also easier to light and burn.

## Stage 6: Re-surveying the site

As part of any clearance programme, it is essential to plan for and introduce a system of re-survey to ensure that any future operations are done promptly and the benefits of clearance and herbicide application are not lost. Stumps or bushes that are recovering from the initial herbicide application (Figure 36), or were missed, need to be identified and a follow-up application programmed into the management plan. Any new seedlings or reinvasion of seedlings from adjacent seed sources need to be located and treated appropriately before the site becomes fully reoccupied (Figure 37).

**Figure 36**

Foliar regrowth from stumps showing partial recovery of foliage to the herbicide treatment.

**Figure 37**

If sites are not successfully treated bush regrowth from live stumps will be rapid and re-occupation of the site by rhododendron will occur.

The re-survey will be based on the initial full site survey, and should only take a day or less to complete per treated site. Each area needs to be surveyed at the end of the growing season following herbicide treatment (i.e. September–October), and should continue until the site is clear of live rhododendron bushes and no longer under threat of reinvasion from adjacent sites.

The re-survey should provide information on a) the **success of herbicide applications**, to trigger follow-up treatments at the appropriate time, and b) identifying those sites that are free from rhododendron but are **liable to reinvasion** if conditions alter.

- **Successful herbicide applications** are those that achieve bush death following treatment to either cut stumps or live foliage. Not all herbicide applications lead to complete stump or bush death; stumps can be missed, or rain can wash herbicide from treated foliage before sufficient herbicide has entered the plant's system.

  Use the map from the initial full site survey (Figure 16) to walk the treated site, noting the extent of stump and/or bush recovery and the density or abundance of stumps and bushes. Recovering stumps or bushes can be allocated to a condition class using Table 6. This table also indicates the follow-on treatment and timeframe for completion if full eradication is to be achieved and the order of priority for each site to be re-treated. This information will allow planning of follow-up treatments, and costing of operations.

- **Reinvasion onto new or recently cleared sites** (Figure 38) is dependent on several conditions being met, principally there being a seed source within dispersal distance of a receptive site (see Seed dispersal, page 4). From the initial survey map the distance to the nearest seed source can be calculated. Assess the potential receptivity of sites that are within 20 m of a seed source within woodlands, and 100 m in open habitats (see Figure 17). Sites of high potential for reinvasion should be re-surveyed at 5-year intervals for regenerating seedlings. Hand-pulling or foliar spraying of new seedlings can then be programmed into the management plan.

**Table 6**

Condition of rhododendron stumps and bushes recorded during the re-survey and the recommended task and timeframe to achieve full eradication and suggested priority.

| Condition class | Follow-up treatment | Timeframe | Priority |
|---|---|---|---|
| Treated stumps | | | |
| Dead (no buds or regrowth, cambial layer under bark dead) | Re-survey | Within 24 months | Low |
| Recovering (live buds present, cambial layer under bark live) | Plan foliar regrowth application (page 20) | Within 24 months | Low |
| Fully recovered with live shoots (regrowth > 0.8 m) | Repeat foliar regrowth application (page 20) | Within 6 months | High |
| Bush foliage | | | |
| Dead (no live terminal buds or cambial layer on shoots) | Re-survey | Within 24 months | Low |
| Deteriorating (continued chlorosis/necrosis and defoliation of leaves) | Re-survey | Within 12 months | Low |
| Healthy and stems <1.3 m in height | Repeat overall foliar herbicide application (page 22) | Within 6 months | High |
| Healthy and stems >1.3 m in height | Cut or flail before applying herbicide as cut stump or foliar regrowth applications (pages 18 and 20) | Within 24 months | Medium |
| New seedlings | Consider hand-pulling or foliar application (page 21) | Within 12 months | Low |

**Figure 38**

Mature rhododendron bushes in this oakwood are acting as the seed source for invasion of new bushes onto the heathland area in the foreground.

# Stage 7: Follow-up control treatments

It is imperative that capacity for follow-up control operations are provided for in the management plan. It is highly unlikely that complete eradication of all live rhododendron bushes will be achieved in a single operation. If the appropriate control technique has been correctly applied, then follow-up treatments are only likely to be required on:

- cut or flailed stumps with regrowth (Figure 39);
- small bushes where ineffective foliar applications were applied (Figure 40);
- seedlings developing on the cleared site;
- cut or flailed branch material that has re-rooted.

Use the results from the survey in conjunction with Table 6 to decide upon the most appropriate application treatment for the type of growth or plant now occupying the site, and carry out the treatment. Further monitoring and follow-up control may be required on some sites for several years, until there is no more live material.

**Figure 39**

An area adjacent to location in Figure 30 four years after being treated by flailing. The untreated bushes (right) are now >2 m tall, flowering annually and as dense as they were before flail clearance.

**Figure 40**

Only the portion of this bush treated with herbicide has been killed; live leaves persist on stems missed by the operators.

# Controlling rhododendron in woodland areas infected with *Phytophthora*

Rhododendron is the main woodland host for two fungal pathogens that have recently been identified in Britain (Figure 41). The pathogens *Phytophthora ramorum* (the cause of Sudden Oak Death in the USA) and *Phytophthora kernoviae*, a newly identified species, are subject to Plant Health regulation. Both species have been found in natural and semi-natural environments in a few discrete locations in southern and south-west Britain causing lethal cankers on a number of oak and beech trees. Diseased rhododendron are a source of fungal spores.

**Figure 41**

Rhododendron acts as the host plant to both *Phytophthora ramorum* and *Phytophthora kernoviae*. (a) *P. kernoviae* infected rhododendron bush. (b) *P. kernoviae* appears to be more virulent than *P. ramorum*, heavily infesting and occasionally killing rhododendron plants.

The symptoms caused by these pathogens on rhododendrons and on trees are described in the plant health pages on the Forestry Commission and Forest Research websites (www.forestry.gov.uk/planthealth and www.forestresearch.gov.uk). To prevent their spread, eradication of rhododendron may be carried out under the direction of the Forestry Commission or other Plant Health Authority. This section describes the best practice for such phytosanitary control measures.

Because of the risk of spreading disease, it is likely that special restrictions will be imposed on the disposal of cut material. These may include burning it all on site or removal in closed containers for safe disposal elsewhere. Disposal routes and methods will be agreed in advance with the appropriate environmental regulator. Washing to remove mud and debris, and disinfection of footwear, clothing and equipment will also be required before moving off-site. All such special precautions will be imposed by a Statutory Notice issued by the Forestry Commission or other plant health authority.

There are four stages for the removal of rhododendron bushes on sites infected with *P. ramorum* and *P. kernoviae*:

1. Remove rhododendron bush top growth.
2. Apply approved herbicide to cut stumps.

3. Apply recommended herbicide to foliar regrowth.
4. Monitor subsequent regrowth or seedling establishment.

## 1. Remove rhododendron bush top growth

Bushes must be cut down with either clearing saws or chainsaws; the risk of spreading infection in debris precludes the use of mechanical flails. Cut the stems so that the stump sits relatively high above ground level to facilitate later operations, but do not leave any live branches or shoots attached to it after cutting. Taller stumps will be easier to find later for herbicide applications and will reduce the chance for reinfection of any shoots that may begin to regrow.

## 2. Apply approved herbicide to cut stumps

Cut stumps will regrow if not treated with herbicide and the new shoot growth is liable to become reinfected with the pathogen: young shoots regrowing close to ground level appear to be more readily infected than older live material. Cut stumps must therefore be treated with an approved herbicide.

The two herbicides approved for use on rhododendron as cut stump applications within woodland are glyphosate (e.g. Roundup ProBiactive) and triclopyr (e.g. Timbrel). Several glyphosate based products marketed by different companies are approved for use in forestry situations. Always check the product label and ensure the product has approval for use in forestry before purchase or application. Herbicides should be applied at the correct product rate (Table 7).

**Table 7**

Rate of product application and best time of year for the treatment of freshly cut rhododendron stumps. Only the herbicides listed are approved for cut stump applications to rhododendron in woodlands.

| Herbicide | Product rate | Optimal time of year |
|---|---|---|
| Glyphosate (360 g $l^{-1}$) e.g. Roundup ProBiactive | 20% solution in water | October–February |
| Triclopyr (480 g $l^{-1}$) e.g. Timbrel | 8% solution in water | October–March |

For safety information on using herbicides consult the Defra publication *Pesticides: code of practice for using plant protection products*, AFAG leaflet 202 *Application of pesticides by hand-held equipment* and the HSE leaflet *Pesticides: use them safely.* Always read the label before applying any herbicides.

## 3. Apply approved herbicide to foliar regrowth

Cut stump applications to rhododendron are rarely 100% successful. Stumps are often missed and not treated, particularly if marker dyes are not used, or the herbicide is not fully effective at preventing shoot regrowth. Furthermore small seedlings are often overlooked and missed in any cut stump treatment. It is vital that a follow-up application of herbicide to any shoot regrowth or small seedlings is planned into the management of control sites.

Three herbicides are approved for directed foliar spray applications to control rhododendron within woodland at the rates and times specified (Table 8). These applications will also kill any live vegetation growing next to or under the target foliage so care must be taken to ensure that only the rhododendron foliage is treated.

**Table 8**

Rate of product application and best time of year for overall foliar applications of herbicides. Only the herbicides listed are approved for foliar applications to rhododendron in woodlands.

| Herbicide | Product rate | Optimal time of year |
|---|---|---|
| 2,4-D/dicamba/triclopyr (200: 85: 65 g $l^{-1}$) e.g. Nufarm NuShot | 7.5% solution in water (maximum of 5 l $ha^{-1}$) | July–September |
| Glyphosate (360 g $l^{-1}$) e.g. Roundup ProBiactive | 2% solution in water plus 2% adjuvant* (High Trees Mixture B) (maximum of 10 l $ha^{-1}$) | March–October |
| Triclopyr (480 g $l^{-1}$) e.g. Timbrel | 2.5 % solution in water (maximum of 8 l $ha^{-1}$) | June–September |

*When near water do not use adjuvant – apply 2.5% solution of glyphosate instead.
For safety information on using herbicides consult the Defra publication *Pesticides: code of practice for using plant protection products*, AFAG leaflet 202 *Application of pesticides by hand-held equipment*, and the HSE leaflet *Pesticides: use them safely*.
Always read the label before applying any herbicides.

## 4. Monitor subsequent regrowth or seedling establishment

Shoot regrowth or new seedlings can become infected with the *Phytophthora* pathogens after the initial clearance operations have been completed. It is therefore vital that the sites are monitored at yearly intervals for at least five years. Recovery of partially controlled stumps, or the establishment of new plants from seed, will only become apparent several years after control measures were started. Any new foliage from either source must be treated with a recommended herbicide before it reaches a height of 1.3 m.

# References and useful sources of information


COLAK, A.H. (1997).
Investigations on the silvicultural characteristics of *Rhododendron ponticum* L. PhD thesis. University of Istanbul, Istanbul.

CROSS, J.R. (1973).
The ecology and control of *Rhododendron ponticum* L. PhD thesis.
University of Dublin, Dublin.

CROSS, J.R. (1975).
Biological flora of the British Isles: *Rhododendron ponticum*.
*Journal of Ecology* **63**, 345–364.

CROSS, J.R. (1981).
The establishment of *Rhododendron ponticum* in the Killarney Oakwoods, southwest Ireland. *Journal of Ecology* **69**, 807–824.

DEFRA (2006).
*Pesticides: code of practice for using plant protection products.* Defra, London.

ERFMEIER, A. and BRUELHEIDE, H. (2004).
Comparison of native and invasive *R. ponticum* populations: growth, reproduction and morphology under field conditions. *Flora* **199**, 120–133.

HEALTH AND SAFETY EXECUTIVE (1997).
*Pesticides: use them safely.* HSE, London.

HEALTH AND SAFETY EXECUTIVE (2003).
*Application of pesticides by hand-held equipment.* AFAG Leaflet 202. HSE, London.

MILNE, R.I. (1997).
Molecular systematics of *Rhododendron ponticum* L. and its close allies. PhD thesis. University of St Andrews, St Andrews.

MILNE, R.I. and ABBOTT, R.J. (2000).
Origin and evolution of invasive naturalized material of *Rhododendron ponticum* L. in the British Isles. *Molecular Ecology* **9**, 541–556.

ROTHERHAM, I.D. (1986).
The introduction, spread and current distribution of *Rhododendron ponticum* in the Peak district and Sheffield area. *Naturalist* **111**, 61–66.

ROTHERHAM, I.D. (1990).
Factors facilitating invasion by *Rhododendron ponticum*. In: *The biology and control of invasive plants*. Proceedings of British Ecological Society Conference. British Ecological Society, Cardiff.

ROTHERHAM, I.D and READ, D.J. (1988).
Aspects of the ecology of *Rhododendron ponticum* with reference to its competitive and invasive properties. *Aspects of Applied Biology* **16**, 327–335.

SHAW, M.W. (1984).
*Rhododendron ponticum* – ecological reasons for the success of an alien species in Britain and features that may exist in its control. *Aspects of Applied Biology* **5**, 231–242.

STEPHENSON, C.M., MACKENZIE, M., EDWARDS, C. and TRAVIS, J.M.J. (2006).
Modelling establishment probabilities of an exotic plant, *Rhododendron ponticum*, invading a heterogeneous, woodland landscape using logistic regression with spatial autocorrelation.
*Ecological Modelling* **193**(3–4), 747–758.

THOMSON, A.G., RADFORD, G.L., NORRIS, D.A. and GOOD, J.E.G. (1993).
Factors affecting the distribution and spread of Rhododendron in North Wales. *Journal of Environmental Management* **39**, 199–212.

WILLOUGHBY, I. and DEWAR, J. (1995).
*The use of herbicides in the forest.*
Forestry Commission Field Book 8.
Forestry Commission, Edinburgh.

WILLOUGHBY, I., EVANS, H., GIBBS, J., PEPPER, H., GREGORY, S., DEWAR, J., NISBET, T., PRATT, J., MCKAY, H., SIDDONS, R., MAYLE, B., HERITAGE, S., FERRIS, R. and TROUT, R. (2004).
*Reducing pesticide use in forestry.*
Forestry Commission Practice Guide.
Forestry Commission, Edinburgh.


## Websites

For further information and details of research on managing rhododendron, visit:

www.forestry.gov.uk/planthealth
www.forestresearch.gov.uk
www.defra.gov.uk/planth/pramorum.htm

For detailed advice on the use of herbicides and appropriate health and safety guidance, visit:

www.psd.gov.uk
www.hse.gov.uk

## Appendix 1

# Application of herbicide by stem treatment

1. Choose a position on the stem as close to the main root system as possible and at least below the last fork. Glyphosate is not readily transported around the bush, therefore only branches above an application point will be controlled, but not those on adjacent stems.

2. Use a heavy-bladed implement or drill a vertical hole in the stem to create a reservoir that will contain all the herbicide solution. Drill bit size should be 11–16 mm diameter (sold as carpentry chisel bits in most hardware stores). Small diameter bits can be used to treat smaller stems, but the larger bit sizes have the advantage of allowing more herbicide contact with the cambial layer. The hole needs to be deep enough to contain 2 ml of herbicide on a level stem (the application of 2 ml of herbicide solution per 7 cm stem diameter is recommended).

3. Apply 2 ml of herbicide to the hole or reservoir immediately after drilling. Take care to prevent herbicide overflowing the hole and running down the stem. The use of a forestry spot gun with a calibrated 10 ml chamber is advised. This allows for the accurate application of a calibrated 2 ml of herbicide per hole. The gun has a 5 l pack that is sufficient to treat 2500 stems. One operator doing both the drilling and herbicide application is more efficient than separate operators and will lead to fewer bushes being missed. Plugging or covering the hole after application is not necessary.

4. Marking treated bushes is advisable. If the operation is in two phases (i.e. one operator drilling holes and a second applying herbicide) spray coloured paint or attach coloured bio-degradable tape to indicate which bushes have been treated. This should also be considered in areas where several operators are working, or if the area is too large to treat in one day.

5. Applications can be made in light rain provided water is not running down the stem into the application hole and washing herbicide solution out into the surrounding area.

6. Bush death should be complete in 9 months from application, but this will depend on the date of initial application. Leaf death can begin within 4–6 weeks following herbicide application in spring or summer treatments. Complete defoliation and bush death usually occur within 9 months on small bushes, and 31 months on those of stem diameters >10 cm.

A suitable stem for injecting herbicide.

A hand-held axe, hatchet or other heavy-bladed implement can be used instead of a drill to create a suitable reservoir for containing the herbicide solution.

Applying herbicide solution into a reservoir with a forestry spot gun fitted with a solid cone nozzle. The gun is calibrated to deliver 2 ml of solution on each application.